REGENERATION

Regrowing Heads, Tails, and Legs

Enslow Publishing
101 W. 23rd Street
Suite 240
New York, NY 10011
USA
enslow.com

Avery Elizabeth Hurt
and Susan K. Mitchell

These books are dedicated to Emily, who inspired the author.
—Susan K. Mitchell

Published in 2020 by Enslow Publishing, LLC.
101 W. 23rd Street, Suite 240, New York, NY 10011

Library of Congress Cataloging-in-Publication Data

Names: Hurt, Avery Elizabeth, author. | Mitchell, Susan K., author.
Title: Regeneration : regrowing heads, tails, and legs / Avery Elizabeth Hurt and Susan K. Mitchell.
Description: New York : Enslow Publishing, 2020. | Series: Animal defense! | Audience: Grade 3-6. | Includes bibliographical references and index.
Identifiers: LCCN 2018051067| ISBN 9781978507180 (library bound) | ISBN 9781978508170 (paperback)
Subjects: LCSH: Regeneration (Biology)—Juvenile literature. | Animal defenses—Juvenile literature.
Classification: LCC QP90.2 .H87 2020 | DDC 571.8/89—dc23
LC record available at https://lccn.loc.gov/2018051067

Printed in the United States of America

To Our Readers: We have done our best to make sure all website addresses in this book were active and appropriate when we went to press. However, the author and the publisher have no control over and assume no liability for the material available on those websites or on any websites they may link to. Any comments or suggestions can be sent by email to customerservice@enslow.com.

Portions of this book originally appeared in *Animal Body-Part Regenerators: Growing New Heads, Tails, and Legs.*

Photo Credits: Cover, p. 1 (axolotl) Lapis2380/Shutterstock.com; cover, p. 1 (sea anemone) Rostislav Ageev/Shutterstock.com; p. 6 AlejandroCarnicero/Shutterstock.com; p. 9 Designua/Shutterstock.com; p. 10 Damsea/Shutterstock.com; p. 14 Nerthuz/Shutterstock .com; p. 17 Joseph T. Collins/Science Source; p. 20 Simon Shim/Shutterstock.com; p. 22 Victor Saul/Shutterstock.com; p. 25 zaferkizilkaya/Shutterstock.com; p. 27 Jeff Rotman/ Oxford Scientific/Getty Images; p. 29 Tiago Sa Brito/Shutterstock.com; p. 31 Jupiterimages/ Photos.com/Getty Images Plus/Getty Images; p. 33 Richelle Cloutier/Shutterstock.com; p. 36 Aurelian Nedelcu/Shutterstock.com; p. 38 Dave Pressland/Corbis NX/Getty Images; p. 41 Bjorn Forsberg/Photolibrary/Getty Images; p. 45 Claus Lunau/Science Source.

Contents

Introduction

For animals, life in the wild can be dangerous. They have to find a place to live and protect their territory. Then they have to find mates. Sometimes, they have to fight for mates. When they have young, they have to protect them—and themselves—from predators. This takes skill.

Over many years, animals have **evolved** in different ways to survive. When it comes to defending themselves from predators, animals have many interesting tricks.

Animals have different defenses depending on the skills their predators have. Rabbits and deer are often preyed on by fast-moving animals, such as big cats. So they have developed the ability to run very fast. This helps them get away from these predators.

Introduction

Turtles are often preyed upon by animals with sharp teeth. They have developed hard shells to protect their bodies. Porcupines have prickly spines they can poke into predators.

Small animals often can't do much to fight off big animals trying to eat them. But fighting isn't always the best idea. Some small animals have developed ways to defend themselves that don't involve fighting. Some frogs have poisonous skin. Skunks can put out a terrible stink. This drives predators away. Some animals use **camouflage** to help them blend in. Predators won't even see them.

These are all cool defenses. But one animal defense is truly amazing. Some animals can regrow lost body parts. This is called regeneration.

Losing a limb is one of the most serious injuries most people can imagine. When people lose limbs, their lives are changed forever. It's the same way for most animals, too. If a fox loses a leg, it won't be able to get around to look for food. It won't be able to run from predators. If it doesn't die from the injury, it might starve to death or be eaten.

But for some animals, losing a leg or a tail is no big deal. They can easily grow another one. This can be a lifesaver.

This lizard is growing a new tail to replace one that it lost, probably while escaping a predator.

First, being able to give up the leg or tail means the animal gets away and doesn't become lunch. Second, being able to regrow the missing body part means the animal doesn't have to try to survive without an important piece of equipment.

Whether these animals live on land, in water, or underground, regeneration is one of the most amazing defenses they use to protect themselves from predators.

Chapter 1

The Special Few

Snap! A predator strikes. It grabs its prey by the tail. The predator is in for a surprise, though. Dinner wrestles free. It runs away leaving only its tail. This is a measly snack for the predator hoping for a full meal. But it works out great for the prey. It gets to live another day. And it may not have to live without its tail. It may be one of the special animals that can grow a new tail! Others can even grow new legs! A few can do a little bit more. They can grow back new eyes or organs. There is even a small group of animals that can grow whole new bodies!

Scientists are not completely sure how the process works. What they do know is that regeneration is possible because of changes in an animal's **cells**. Cells are the smallest basic building block of every living thing.

There are many kinds of cells. Cells that are alike gather together to form different parts of the body. Some cells make up skin. Others form bone or muscle. There are special cells in nerves and blood. Each kind of cell has a different job in an animal's body. In most animals, most of the time, cells can't change and do a different job. A skin cell is always a skin cell. A blood cell is always a blood cell. If you need more blood cells, skin cells won't be able to help.

A regenerator's cells are special. These cells can change jobs. The different cells in the spot where a leg or tail is lost, for example, stop doing their own special jobs. Instead, cells that had different jobs all start doing the same thing. They all help to regrow the missing body part.

Show Some Spine

Almost all animals that can grow new body parts are **invertebrates**. That means they do not have backbones, or spines. They also do not have a skeleton inside their bodies. Instead, some of them wear a hard outer skin called an **exoskeleton** that completely covers their

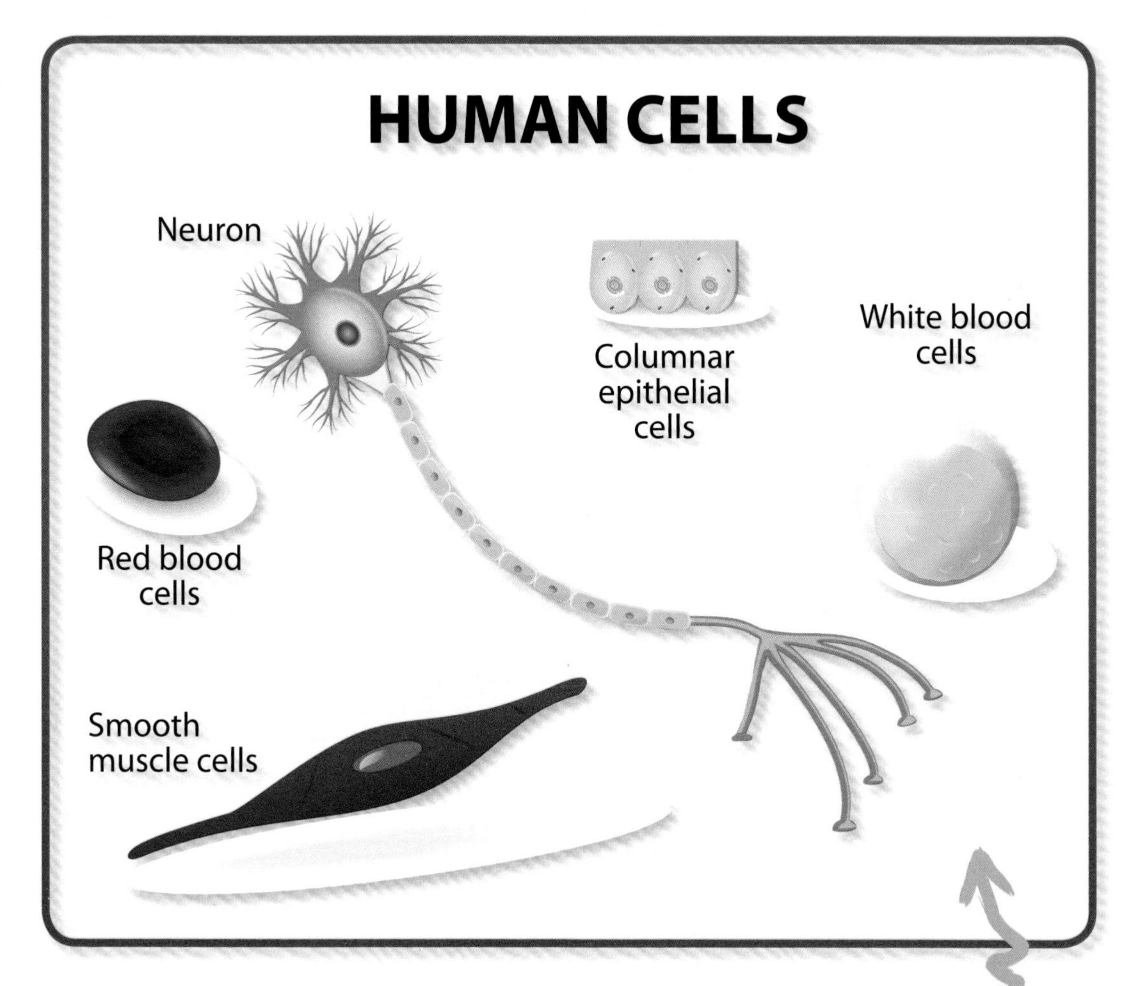

Mammals have many kinds of cells. These are a few types of cells in human bodies.

bodies. Others simply have soft bodies or thick shells that partly cover them.

Insects, worms, and spiders are all invertebrates. Animals such as clams, starfish, and crabs have no backbone, either. Many animals in this group have very simple bodies. It is easy to forget that they are even animals!

Sponges are a good example of super simple animals. They have no true organs. They are not much more

This red sponge is called a boring sponge. It's not uninteresting. It's called a boring sponge because it bores holes in oyster shells.

than a group of living cells. Sponges come in many shapes, sizes, and colors. One thing all sponges have in common are their pores. Pores are the holes all over a sponge's body.

These holes let water pass through the sponge. Food is filtered from the water. Sponges cannot move, so they simply sit and wait for food. This means that they cannot get away from predators. However, a sponge can regenerate any part of its body that might get eaten. So unless an animal eats the entire sponge, it hasn't done much harm.

What Nerve!

Animals that do have skeletons and backbones are called **vertebrates**. All birds and mammals have backbones. Reptiles such as snakes and lizards also belong to this group. So do frogs and their relatives. Although vertebrates have some power of regeneration, very few can grow new body parts.

Animals with backbones usually have complex nervous systems. The nervous system controls the body. It usually includes a brain and spinal cord. It also has a huge network of nerves. This system controls everything the body does, including movement and breathing. It controls muscles. It also allows animals to feel temperature and pain.

Most invertebrates have nervous systems, too. Their nervous systems are much simpler, however.

The Swiss Are Stunning

In 2018, the government of Switzerland made it against the law to put living lobsters in boiling water. Dropping a live lobster into boiling water is the most common way of cooking the animal. However, most scientists think lobsters suffer when they are killed this way.

Some scientists disagree, though. They say that lobsters may not feel pain the way other animals do. Their brains don't have the parts other animals use to process pain. But lobster brains are very different, so scientists aren't sure. Lobsters may have other ways of feeling pain. The Swiss decided it was best to be careful. The Swiss government recommends killing lobsters by stunning them. It's much more humane.

They usually have very simple brains. Even with simple brains, invertebrates are able to take in a lot of information. They respond to being touched and to changes in light and temperature.

Invertebrates do not have spinal cords either. They do have a simple network of nerves. That means most invertebrates can feel pain. Losing a leg, tail, or other body part can really hurt. It also takes energy to regrow a lost limb. Some animals can regrow limbs fairly quickly, but it still takes some time. For all these reasons, most animals will use many other defenses before they leave a leg behind.

Gone for Good

Mammals are not able to grow new body parts. Since humans are mammals, that is true for us, too. If a person loses an arm or leg, it is gone for good. The wound may heal, but the missing part will never grow back. Humans can regenerate some things, though. For example, new skin can grow to heal a wound. Bones can grow back together after breaking. New blood cells are made to replace lost blood.

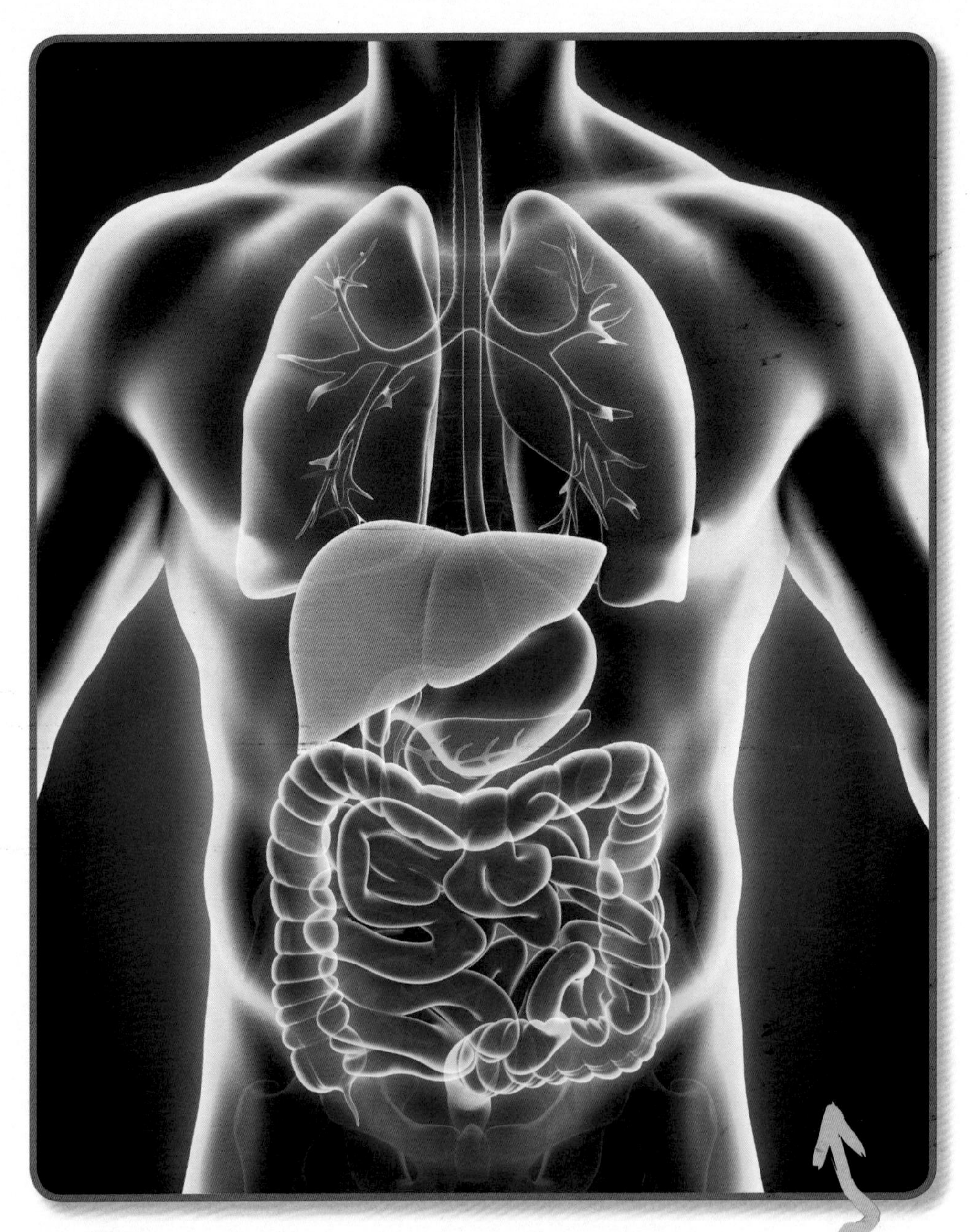

Even humans can regenerate some parts of themselves, such as liver cells.

Fun Fact!

Most invertebrates are small, but not all. The largest invertebrate in the world is the colossal squid. It can weigh 1,000 pounds (454 kilograms) and be almost 33 feet (10 meters) long.

The human liver has amazing regenerating ability. It is an organ in the digestive system and does many important things. If the liver is damaged, doctors may remove part of it. The part of the liver that is left will grow. It is one of the only parts of the human body with this power.

Chapter 2

In and Out of the Water

Salamanders and newts have many different defenses. Some use camouflage to blend in with their surroundings so that predators can't see them. They also might try to act vicious or look big to scare off a predator. Some have poisons in their skin. Others can make their ribs poke through their skin. This could be a sharp and unpleasant surprise for an animal planning to eat a salamander. But no defense is perfect. Salamanders and newts often get caught by other animals. Fortunately, they also can survive the loss of a tail or leg, thanks to their super regenerating powers.

Salamanders and newts, like frogs, are **amphibians**. A newt is a kind of salamander, but there are some

This salamander is just starting to regenerate a lost tail.

differences between the two animals. Newts usually have rougher skin than salamanders. They also return to water as adults, while salamanders stay on land for the rest of their lives. These animals all begin life in the water. They also go through different stages in their life cycle. A salamander or newt starts out as a small, gooey egg. After hatching, it enters a larval stage. It looks like a small blob with a tail but no legs. At this stage in life, it is able to grow back body parts quicker and more easily than it can as an adult. But even as an adult, it still has some ability to regenerate.

Fun Fact!

Global warming is causing some species of salamanders to shrink. They have to use more of their energy to stay cool as temperatures get hotter. This leaves less energy for growing.

Fast Healing

While they are living in the water, the biggest threats to salamander and newt larvae are fish. Once on land, the adults face many new predators, such as birds, snakes, raccoons, and even other amphibians such as frogs. Most of the time, salamanders and newts stay hidden under rocks and leaves, away from predators.

Both newts and salamanders are predators, too. They are carnivores, which means that they eat other animals. Their favorite foods are worms and insects. Eventually, newts and salamanders have to come out of hiding to hunt for food. That is when they are in the most danger.

While out hunting, a newt or salamander is mostly unprotected. It could easily be caught by another animal. Like most animals, the newt or salamander tries very hard to protect its head. If a predator catches it by the head or neck, it will most likely not survive. On the other hand, if a predator grabs a newt or salamander by the tail or legs, there is still hope for escape. Salamanders and newts can repair the damage very quickly. In less than twelve hours, a new leg begins to grow and replace the missing one.

Frogs also have some power to grow new body parts, but only when they are tadpoles. A tadpole is a frog's larval form and looks nothing like a frog. It is a legless black blob with a tail. As the tadpole grows, legs start to sprout and the tail starts to shrink. If a predator breaks off the tadpole's tail or leg, the tadpole can grow another one to replace it.

A tadpole changes from a blob to a frog by sprouting legs and losing its tail.

Once a tadpole has grown into an adult frog, it loses the ability to regenerate lost body parts. Fortunately, for salamanders and newts, they keep this amazing ability all their lives.

Lizards have regeneration powers, too. Getting caught by a predator is never a good thing. But if a lizard gets caught by the tail, at least it has a chance of surviving. That is because a lizard can lose its tail without serious damage. There are limits to lizard regeneration, however. Some lizards can grow back only a small bit of the lost tail. Others may be able to grow it back only once. But the fact that a lizard can lose its tail is actually more important than the ability to grow it back. While a predator is busy with a wiggling lizard tail, the lizard can run to safety!

Under the Microscope

The secret to the newt's and salamander's regeneration is in their cells. The cells can take over the jobs of other

Look-Alikes

Salamanders and lizards look a lot alike. But they are completely different kinds of animals. Salamanders are amphibians. Lizards are reptiles. Reptiles are usually covered in dry skin. Amphibians have moist skin. Amphibians have to lay their eggs in water or a very moist place. Lizards can lay their eggs on dry land. Another way to tell lizards from salamanders is to look for ears. Lizards usually have small ear openings. Salamanders do not. Lizards usually have claws on their toes. Salamanders don't have these either. But both animals have at least some powers of regeneration.

cells. This allows the newt or salamander to grow new skin. It also means that they can grow new muscle and bone. The cells begin to grow and divide. Two cells split and become four, and so on. Soon, there are enough cells to create a new body part.

The axolotl might be able to help scientists learn how to heal people who have had spinal cord injuries.

In and Out of the Water

In most animals with a backbone, any body part that grows back will not be the same as the first one. Arms, legs, or tails may not be as big or as long. The salamander and newt, however, can grow an exact copy of the original limb. At first glance, no one would ever know the part had ever been missing.

Scientists still do not know how newts and salamanders can do this. To learn more, scientists study these animals in a research lab. By watching the salamanders repair new legs, tails, and even eyes, they hope to figure out exactly how it is done. Salamanders and newts may one day hold the key to helping scientists learn how to help humans who have lost limbs.

One of the most popular salamanders used for research is the axolotl (AK-sa-lot-el). It is found in central Mexico. It is not only able to regenerate a lost tail or legs, it can regrow parts of its heart. It can grow new jaws. It can even grow new lenses for its eyes. Most amazing is its ability to regrow parts of its spinal cord!

Chapter 3

Deep-sea Defense

Many people have probably read about **clones** in comic books or seen them in movies. But clones exist naturally in the animal world. Some animals can clone themselves with no help from scientists. In real life, clones are a little bit different than in fiction. Clones in nature are a bit different from clones in the laboratory. And they are even more exciting.

A clone is an exact copy of a living thing. It is grown and created from the cells of an animal or plant. Starfish are a great example of an animal that uses cloning for defense. They are the masters of regeneration. Through regeneration, they don't just replace lost limbs. They can make exact copies of themselves.

Deep-sea Defense

An entirely new starfish body is regenerating from just an arm.

Despite their name, starfish are not fish. They belong to a group of animals called **echinoderms** (a-KINE-a-derms). Sea urchins are also echinoderms. The word "echinoderm" means spiny-skinned. While most starfish don't have large spines or spikes like sea urchins, they do have rough or bumpy skin.

A starfish's tough body wall keeps most animals away, but some ocean predators, such as sharks and rays, like to eat starfish. Some snails and shrimp also eat them. The biggest threats to starfish, however, are other larger starfish!

Fun Fact!

Because they are not fish, some scientists prefer to call starfish sea stars.

There are more than one thousand kinds of starfish, in every size and color. Starfish live in most of the oceans in the world. They can live in warm, tropical seas and in icy polar oceans. They can live in shallow tide pools or on deep, dark ocean floors. Starfish can live just about anywhere.

All starfish have more or less a star-shaped body. Most are easy to spot by this five-armed star shape. But some starfish do not stick to the five-arm rule. The sunflower starfish can have as many as twenty-four arms. The crown-of-thorns starfish can also have more than five arms. The cushion starfish looks like a pentagon (a five-sided shape). It has no arms at all.

Well Armed

A starfish's arms help it do many things. They help the starfish move from place to place. They also help it eat. Underneath the arms are tiny tube feet. Tube feet look like very small tentacles. There can be hundreds of tube feet under each starfish arm.

The starfish controls its tube feet with small muscles. These muscles pull water from inside the starfish's body and force it through the tube feet. This helps push the starfish along the ocean floor. Despite having so many tube feet, starfish cannot move fast.

The tube feet on the arm of this sunflower starfish help it move and eat.

They move so slowly that sometimes it looks like they are not moving at all.

To eat, a starfish uses the small suction cups at the end of each of its tube feet. These give the starfish an amazingly strong grip. Starfish can easily pull open a clam or other shelled animal. They also use their tube feet to bring food to their mouths. Some starfish can push their stomachs outside of their mouths to eat!

Starfish may look like they are all arms, but they do have a body. The center part of a starfish is its body. It is called a central disk and is usually divided into five sections. Each section usually has an arm attached.

The center part of a starfish is where its two stomachs are located. The mouth is located underneath. A starfish does not have a heart or brain. Instead, it has a nerve ring around the mouth area. Simple nerves branch off the ring to each arm.

No Big Loss

Most of a starfish's major organs are inside its arms. Each arm has reproductive organs. Each arm also has pouches that come off of the main stomach.

Deep-sea Defense

Starfish arms are made to break off very easily. If a starfish is caught by a predator, it can simply lose its arm. The predator is usually too busy with the arm to notice that the rest of the starfish has gotten away.

If a starfish loses an arm without also losing part of the central disk, it will grow another arm. The new arm is often smaller than the original arm. But it still works like the old one, and the starfish will survive.

On the other hand, if a lost arm includes part of the central disk, the arm itself can also regenerate. The damaged arm will begin to grow four more arms. Soon, there is a completely new starfish from just one arm! By losing its arm, not only will the original starfish

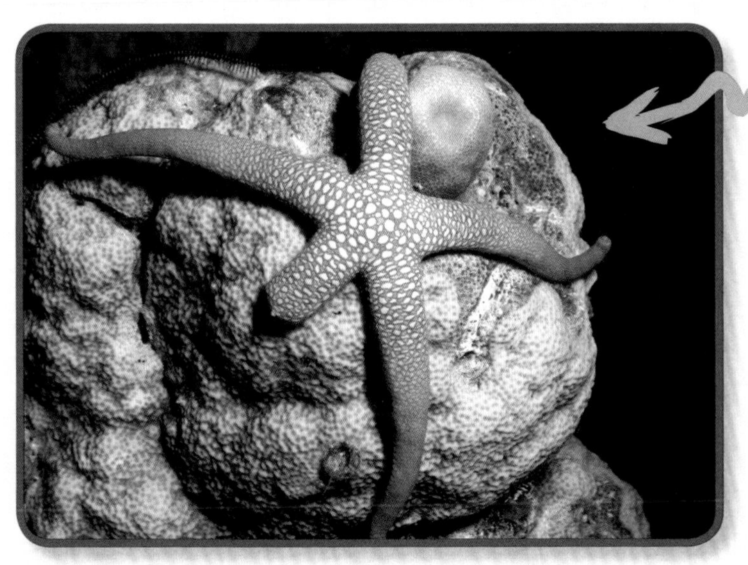

This starfish has lost part of one of its arms. But it can easily grow it back.

live through an attack, but an entire new one might be created.

Regrowing Guts

The sea cucumber is another animal that has tube feet. Like starfish, sea cucumbers are also echinoderms. But sea cucumbers look very different from other echinoderms. These creatures have long, thick bodies.

Revenge of the Clones

The ability of starfish to clone themselves helps protect the starfish population. A long time ago, people who fished for oysters would chop starfish into several pieces before throwing them back into the water. They thought they were killing the starfish. What they were really doing was making several more starfish.

On top of that, scientists have discovered that starfish that reproduce themselves by cloning live longer than those that reproduce the normal way. They've also found that those cloned starfish are healthier.

Deep-sea Defense

Sea cucumbers have a strange—and slightly gross—way of escaping predators. They can replace any organs they lose in the process.

A sea cucumber looks more like a lump on the bottom of the ocean than an animal.

Sea cucumbers have a very strange way of escaping predators. They can push some of their organs out of their bodies! While a predator is distracted by the guts floating freely in the water, the sea cucumber has a chance to get away. Losing their insides would kill most animals. Not only will the sea cucumber survive, it can regenerate the lost organs.

Breaking Up Is Easy to Do

Starfish can regenerate entire new starfish. Most animals can't go anywhere near that far. But it's still useful to be able to replace a leg or two from time to time. Crayfish, also called crawfish, crawdads, or mud bugs, can do this. They also have a special spin on this amazing skill.

A crayfish is one of the few animals that can choose to lose a leg. Most animals that can grow new body parts don't lose limbs on purpose. Their limbs are bitten off or ripped off by predators. The crayfish can make its leg pop off whenever it wants. That is a handy trick if a predator has hold of a crayfish by the leg.

Crayfish look like tiny lobsters. In fact, they are actually a close cousin of the lobster. Both belong to a

This red crayfish looks a lot like its larger cousin, the lobster. Lots of people love to eat crayfish, too.

group of animals called **crustaceans** (krus-TAY-shuns). Shrimp also belong to this group. Crayfish are found in the southern United States, as well as in many other places, such as Asia, Europe, and New Zealand. In North America, there are more than three hundred species of crayfish. There are especially plentiful in the Southeast, where they are called crawfish. Red is the most common color, but there are blue and white crayfish, too.

Most crayfish live in freshwater rivers and streams. Very few crayfish live in saltwater. But crayfish can live in many different areas. Some live in swamps or small ponds. Others are just as happy living in a roadside ditch.

During the day, crayfish spend their time hiding under rocks or logs. They are mostly nocturnal. That means they are active at night. That is when they hunt for insect larvae and small worms or fish. Plants are also on their menu. They will eat just about anything. Eating both plants and animals makes them **omnivores**.

Cracking Up

A crayfish's body is made of two parts. The front part includes the head and is called the cephalothorax

Fun Fact!

After molting, crayfish often eat their old exoskeletons. This way they can reuse the minerals it contains to help make the new skeleton.

(sef-a-low-THOR-ax). The back end of the crayfish is the abdomen. It is made up of several **segments**, or sections. Crayfish are partly covered in a very hard outer shell called a carapace. The carapace is also called an exoskeleton. This helps protect the crayfish's soft body.

The carapace of a crayfish does not grow. When the crayfish's body grows, it must shed the shell. This is called molting. When a crayfish molts, the hard outer shell splits open. The crayfish crawls out of the too-small shell. During this time, it is in a lot of danger. Its body is soft and unprotected. After a few days, a new shell hardens and the crayfish is protected again.

The exoskeleton gives the crayfish some safety, but it is not perfect. Many predators have learned to crack the hard outer shell. Others have teeth that can easily bite through the crayfish's shell. A crayfish needs a defensive backup plan. That is where its legs come in.

The crayfish has four pairs of walking legs. These legs are attached to the cephalothorax.

On the crayfish's abdomen are five pairs of smaller legs. Although these are not true legs, they do help the crayfish move around.

At the top of each leg, the crayfish has a "breaking joint." A **joint** is the place where body parts meet, such as elbows, knees, and shoulders on a human. Joints allow body parts to move. The breaking joints of a crayfish are found right where the eight walking legs meet the cephalothorax.

A crayfish has other weapons it can use against predators. It has two front legs with large claws. These strong claws can give a very painful pinch! These claw arms have a breaking joint, too, and can grow back if they are broken off.

Here, the joints of this crayfish's legs are visible. This female has red eggs in her tail.

How Old Is Your Leg?

Like crayfish, many insects and spiders can grow new legs if they lose the old ones. But this works only if they lose one or two legs. Losing too many legs is a problem.

They're Trying to Tell Us Something

Many animals are endangered because of habitat loss. Human-built dams, erosion, and water pollution make it hard for crayfish to find places to live and breed. But crayfish may be able to help solve at least one of these problems: pollution. Scientists call crayfish a keystone species. A keystone species is a species that all the other animals in a habitat depend on. If a keystone species disappears, the habitat changes dramatically. By keeping track of how many crayfish live in an area, scientists can get a good idea of how much trouble that environment is in.

Crayfish can grow new legs at any age. Spiders, like frogs, can regenerate lost legs only when they are young. A baby spider can grow back a lost leg. An adult spider cannot. Like a crayfish, a spider's outer skin does not grow with it. A spider molts as it grows. Since a young spider will molt several times before it

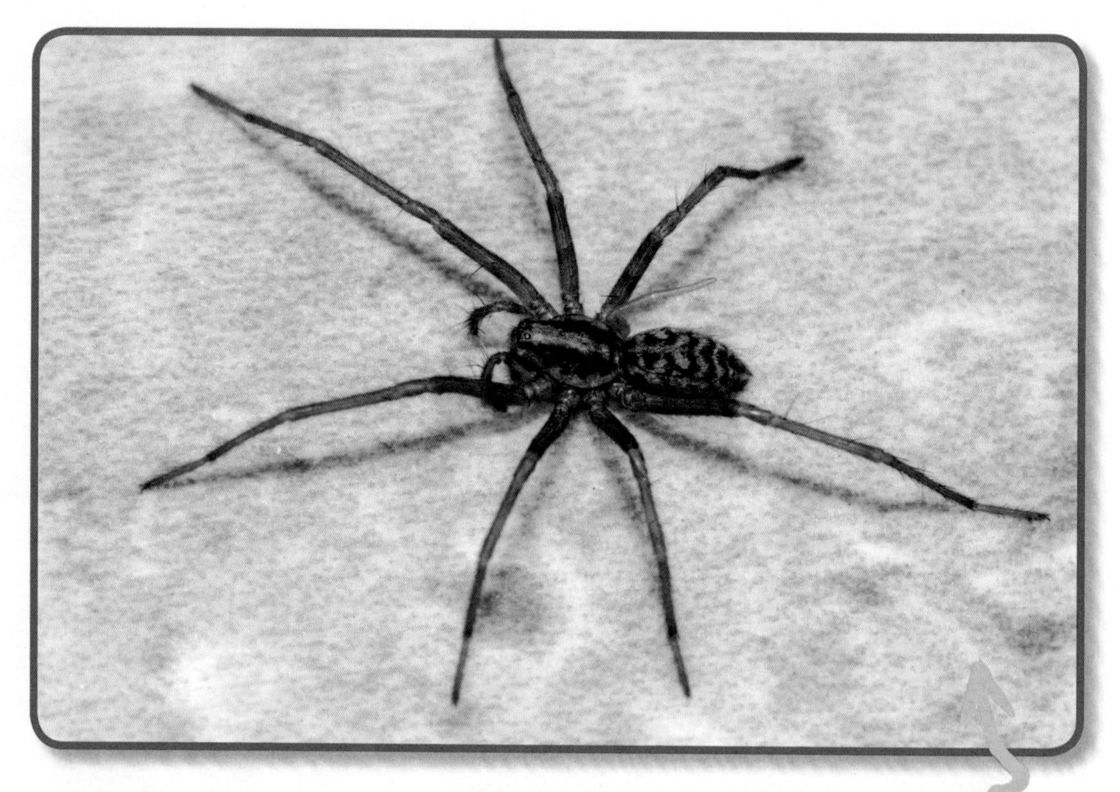

This female dust bunny spider is in the process of regenerating her right back leg. It's lighter in color and smaller than the other legs.

becomes an adult, it has many chances to regenerate if it has to.

A baby spider's new leg will get bigger each time the spider molts. When a spider is fully grown, it stops molting. Then the spider can no longer grow back a lost leg. Sometimes, a spider may lose a leg and have only one or two more molt cycles left in its life. This means a leg might grow back, but it will not be as large as the other legs.

Insects, spiders, and crayfish all use their legs for many reasons. They use them for movement. They can use them for defense. They also use them for getting food. If these animals lose too many legs, they could die before they are able to grow new ones. Without legs that can catch food, they will starve before regeneration is finished.

All animals that can lose a body part on purpose have breaking joints that can easily break away when the animal needs to drop a leg or tail. This also helps make sure that the new leg will be the same size and just as strong as the other legs.

Chapter 5

The Regeneration Underground

Animals on land, in swamps, and in the sea have amazing powers of regeneration. Even some animals that spend most of their lives underground can regenerate body parts. Earthworms live under the soil. What body part does an earthworm have to lose? It doesn't have arms and legs. It turns out that an earthworm can regenerate most of its body!

An earthworm cannot move very fast. It digs through the soil, eating dirt. That might not sound very good to a human, but the dirt is full of minerals and other things that are very healthy for the earthworm.

All of this digging and eating make earthworms very important to plant growth. The tunnels earthworms make under the soil let air pass through the soil and help water flow to plant roots. The dirt that is digested by the worm helps plants as well.

Earthworms are a favorite meal of many animals, including birds, frogs, turtles, weasels, and otters. The simple, dirt-eating life of an earthworm is far from safe.

Coming Up for Air

The first line of defense for any earthworm is to stay hidden. Living under the ground helps, but only a little. Plenty of hungry animals know they're down there. Also, even earthworms can't stay underground all the time.

This bird has found a nice, juicy earthworm to snack on. If the earthworm escapes, it may have to regrow some of its body.

The earthworm does not have lungs. It breathes through its skin. This means that earthworms must always stay moist. If their skin dries out, they cannot breathe and will die. They get air from spaces in the loose soil. But when it rains, these spaces fill up with water and the earthworm has to come to the surface to breathe.

Coming to the surface means the earthworm is out in the open. It makes it easy pickings for a predator. Even if the earthworm stays underground, it is not completely out of danger. Many mammals dig deep to find earthworms to eat.

The earthworm does have a few tricks to stay safe. First of all, an earthworm is an invertebrate. Without a backbone, the earthworm can twist and wriggle in almost any shape or direction. It can be hard for a predator to get a good grip on a wiggling worm.

Also, an earthworm's body is lined with very tiny hairs called setae (SEE-tee) that help it move from place to place. Setae also can act like anchors to keep the worm in one spot. They can stick tight against the walls of the worm's tunnel. This can make it harder for a predator to pull an earthworm out of its hole.

Fun Fact!

Earthworm skin has a slippery liquid on it. This helps the worms move easily through their underground tunnels.

Careful Where You Cut

Sometimes, all of a predator's pulling can cause disaster to strike. An earthworm can be pulled apart. Fortunately, the earthworm has a secret defensive weapon. Even if the predator gets a mouthful of the worm's tail, the earthworm might still survive. That is because an earthworm can regenerate most of its body.

An earthworm's body is made up of many segments. These are ring-like body parts. Some earthworms may have as many as one hundred body segments, or even more. The front segments of an earthworm hold the important body organs. Inside these segments is a very small, simple brain. There are also five hearts. The front segments are also where the earthworm's major digestive organs are located.

Clearly, an earthworm's front segments are the most important. If the earthworm is split, the front end will grow new tail segments. The cut-off tail segments of the injured worm might also regenerate. The only problem is that this will make a worm with two tail ends instead of one with a head end and a tail end. Eventually, the worm with two tail ends will

Lessons from the Wild

Humans sometimes lose limbs in accidents or war. It would be great if people could regrow those limbs. Scientists all over the world have spent years studying animals that can regenerate. Many of them study stem cells. These are special cells that can become any type of body part, not just one. Scientists hope that by figuring out how stem cells work, they can one day find a way to help humans regenerate limbs.

This diagram shows how the important organs are all in the front of the earthworm's long body.

die of starvation.
How an earthworm regenerates all depends on where it is cut in two!

Growing new body parts sounds a little like science fiction. But in the animal world, it is very real. Having the ability to live through a predator's attack that could kill another animal gives regenerators a huge advantage. They can escape, heal, grow, and survive.

Glossary

amphibian A cold-blooded animal, such as a frog or salamander, that can live both on land and in water.

camouflage A defense in which an animal's coloring or shape helps it hide from predators.

cell The basic building block of a living organism.

clone An exact copy of a living thing.

crustacean A member of the class of animals that live in water and have shells.

echinoderm A member of the group of ocean animals with spiny skins, such as the starfish.

evolve To gradually change over time in response to the environment.

exoskeleton The hard outer covering of some animals that provides protection.

habitat The natural environment where an animal lives.

humane Kind or compassionate.

invertebrate An animal with no backbone.

joint A place where two parts of an animal's skeleton come together, allowing for movement.

omnivore An animal that eats both meat and plants.

segment One of several parts into which something is divided.

vertebrate An animal with a backbone.

Further Reading

Books

Hesper, Sam. *Fire Salamanders*. New York, NY: Rosen, 2015.

McAneney, Caitie. *Why Don't Jellyfish Have Brains? And Other Odd Invertebrate Adaptations*. New York, NY: Gareth Stevens, 2019.

Royston, Angela. *Invertebrates*. Portsmouth, NH: Heineman, 2015.

Turner, Matt. *Extraordinary Insects*. Minneapolis, MN: Lerner, 2017.

Websites

National Geographic Kids: Earthworm
kids.nationalgeographic.com/animals/earthworm
Read more about the earthworm.

National Geographic Kids: Mexican Axolotl
kids.nationalgeographic.com/animals/mexican-axolotl
Learn more about the amazing axolotl.

National Geographic Kids: Sea Star
kids.nationalgeographic.com/animals/sea-star
Discover more facts about the starfish.

Index